COUNTDOWN TO SPACE

MOON BASE

First Colony in Space

Michael D. Cole

Series Advisor:
John E. McLeaish
Chief, Public Information Office, retired,
NASA Johnson Space Center

Enslow Publishers, Inc.

44 Fadem Road
Box 699
Springfield, NJ 07081
USA

PO Box 38
Aldershot
Hants GU12 6BP
UK

http://www.enslow.com

Library of Congress Cataloging-in-Publication Data

Cole, Michael D.
Moon base : first colony in space / Michael D. Cole.
p. cm. — (Countdown to space)
Includes bibliographical references and index.
Summary: Describes the *Apollo 11* mission to the moon, explains the need for establishing a moon base, and speculates about future situations in which the base would be used.
ISBN: 0-7660-1118-6
1. Lunar bases—Juvenile literature. 2. Moon—Exploration—Juvenile literature. 3. Space colonies—Juvenile literature. [1. Lunar bases 2. Moon—Exploration. 3. Space colonies.] I. Title. II. Series.
TL799.M6C624 1999
919.9'1'04—DC21 98-13126
CIP
AC

Printed in the United States of America

10 9 8 7 6 5 4 3 2 1

To Our Readers: All Internet addresses in this book were active and appropriate when we went to press. Any comments or suggestions can be sent by e-mail to Comments@enslow.com or to the address on the back cover.
The Publisher

Illustration Credits: National Aeronautics and Space Administration (NASA), pp. 4, 7, 9, 12, 14, 16, 19, 21, 23, 25, 27, 29, 33, 35; NASA/JPL, p. 39; NASA/JPL/CALTECH©, p. 37.

Cover Illustration: NASA(foreground); Raghvendra Sahai and John Trauger (JPL), the WFPC2 science team, NASA, and AURA/STSCI (background).

Cover Photo: Astronauts on the future Moon base load their spacecraft before they return to Earth.

CONTENTS

Future cargo flights to the Moon may use a Russian launch vehicle to carry important materials to the surface.

"*Apollo*, Houston. We're go for undocking," said Mission Control in Houston, Texas.[1] The capsule communicator, or capcom, was speaking to the crew of *Apollo 11*. The crew was more than 230,000 miles away in space, in orbit around the Moon.

Astronauts Neil Armstrong and Edwin "Buzz" Aldrin undocked the lunar landing module *Eagle* from the command module *Columbia*, piloted by astronaut Michael Collins. The two spacecraft flew in formation around the dark side of the Moon until they emerged into the sunlit side again. "The *Eagle* has wings," Armstrong radioed, just before *Eagle* drifted slowly down and away from Collins in *Columbia*. *Columbia* would remain in orbit.

"You are go to continue powered descent," Mission Control said. "You're looking good." The buglike *Eagle*, with its four landing legs, was lowered toward the surface of the Moon.

It was July 20, 1969. Armstrong and Aldrin were headed toward one of the most historic moments in human history. They were going to land on the Moon. All was going well as they descended toward the surface. Then the landing site came into view.

Suddenly there was trouble.

The intended landing site was littered with large boulders! If they lowered the lunar module into that area, they would crash into one of the boulders. Armstrong decided to override the automatic landing system. He grabbed the rocket control handle with his right hand. The landing was now up to him. There was very little fuel left to carry *Eagle* the extra distance over the boulders. They were now only one hundred feet above the Moon.

"Sixty seconds," Mission Control said. There were only sixty seconds worth of fuel in *Eagle*'s landing engine! Armstrong had to find a smooth landing area in a hurry. He saw an area several hundred yards to his right. As he and Aldrin continued to descend, Armstrong carefully maneuvered *Eagle* in that direction.

Aldrin read out to Armstrong, "Picking up some dust. Thirty feet . . . faint shadow . . . drifting to the right a little."

"Thirty seconds," said Mission Control. Armstrong and Aldrin were almost there.

Eagle's engine kicked up lots of dust. Then the dust cloud moved away from them. They had come to a complete stop.

"We copy you down, *Eagle*," said Mission Control.

"Houston," Armstrong said hesitantly, "Tranquility Base here, the *Eagle* has landed."

Human beings, for the first time, had landed on the Moon.

Some hours later Neil Armstrong, in his bulky white space suit, emerged from *Eagle*'s hatch at the top of the landing ladder. Millions of people on Earth

Humans first landed on the Moon in 1969. Soon we may return to begin construction of a permanent base on the Moon.

watched the amazing live television pictures from the Moon.

"Okay, Neil, we can see you coming down the ladder now," Mission Control said. Armstrong made his way carefully down each step until he stood at the base of the ladder. He then lifted his foot and stepped away from *Eagle*.

"That's one small step for a man, one giant leap for mankind," he said.[2]

Eleven other astronauts followed in the footsteps of Neil Armstrong's giant leap. From July 1969 through December 1972, a total of twelve American astronauts walked on the Moon during six Apollo missions. Since then, no humans have returned to the Moon.

In the 1960s there was great support for the National Aeronautics and Space Administration (NASA). A state of tension existed between the United States and Russia, which was then part of the Soviet Union. For many reasons, both countries believed that one way to prove which government was strongest was to become the first country to land its people on the Moon. The government and people of the United States wanted to win the space race against the Soviet Union. For that reason, the United States space program had the strong support of the public.

After the success of the Apollo program, the public's opinion changed. The huge cost of the missions made many Americans question the importance of going to the

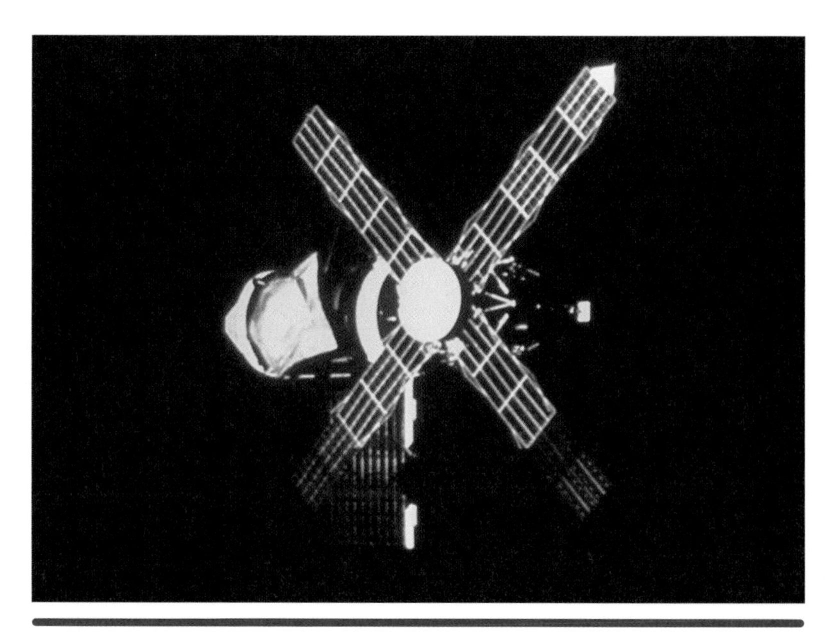

With the Skylab *program in the 1970s, scientists continued to learn about the effects of space on human beings.*

Moon to gather lunar rocks. After the excitement of winning the space race, many Americans lost interest in the space program.

Americans stayed in space with *Skylab* in the mid-1970s and the space shuttle program in the 1980s and 1990s. Russia remained in space with the *Salyut* and *Mir* space stations. But neither Americans nor Russians made any trips to the Moon.

On July 20, 1989, twenty years after the first Moon landing, President George Bush met with the crew of *Apollo 11* to announce the Space Exploration Initiative. Years earlier, President Ronald Reagan had committed

America to building a space station. Now, President Bush extended that commitment to a long-range plan of space exploration that included the Moon and Mars.

"Next," Bush said, "for the new century—back to the Moon, back to the future, and this time, back to stay."[3]

Since the days of the Apollo missions, the machines and technologies for space travel have become better and safer. Spacecraft computers are far more powerful today. Scientists know much more about how the human body reacts to being in space. NASA, the Russian space program, and the European Space Agency are also finding cheaper ways to build and fly their robotic spacecraft. Robotic probes and rovers do not risk human lives and will play a major part in getting back to the Moon.

With the increased knowledge, technology, and interest for returning astronauts to the Moon, it appears the time has finally come.

It is time to build a Moon base.

2
The Next Step

NASA has plans for establishing the first colony in space on the Moon. Building a Moon base is the next step toward humans living away from Earth. It usually takes people a little time to adjust when they move to another neighborhood or city. Imagine the task of learning how to live in a place where there is no atmosphere and no soil in which to grow food! A colony in space will be the first site where humans will live and work on a surface other than Earth's. It will give scientists and engineers an opportunity to test the techniques and equipment required for living away from Earth.

In addition to learning how humans could live away from Earth, scientists could conduct research on the surface of the Moon. A permanent Moon base would

The Moon, our nearest neighbor in space, may someday be the site of a busy space colony.

serve as a place for testing the machines, equipment, and human capabilities that would be needed for many future space missions, including a mission to Mars.[1]

The Moon is the nearest object to us in space. It orbits Earth at approximately 240,000 miles away. If anything ever went terribly wrong on the Moon base, the crews would be, at most, only a few days away from home.

To share the cost of going to the Moon and building a base, the United States, Russia, and the European Space

Agency (ESA) will all likely be involved in the Moon base project.

"We see our Moon program as an invitation to the world," said ESA's Roger Bonnet in 1995. "The program should have the broadest possible consensus and involvement, and could . . . ultimately involve all citizens of Earth." During the Moon base project, ESA wants to provide the public with as much access to the Moon experience as possible. Through television broadcasts and a system of computer-generated simulations using actual images from the Moon, ESA wants the public to experience the lunar activities from home.

"The Moon is already part of human territory," Bonnet added, "and it should be accessible to all humans, not only a select group of astronauts."[2]

There are three phases to the Moon base project. The robotic phase of lunar study represents the first phase. Phase Two is the landing of habitat modules and equipment for a permanent lunar outpost. Astronauts will also be going to the Moon in Phase Two. The modules will provide a comfortable living environment for the astronauts. Phase Three is the expansion of the outpost to a working lunar base, which will gradually lessen its dependence on materials and supplies from Earth.[3]

Even though astronauts have already visited the Moon, they have not been the first to return. Robotic spacecraft and probes went ahead of them to perform detailed studies of the lunar surface. The first lunar probes landed in the 1960s to prepare the way for the Apollo missions. More recently, the orbiting *Clementine* spacecraft surveyed the Moon's south pole in 1994. *Clementine* revealed areas of deep craters that are permanently shadowed. Scientists believed these permanently shaded areas might contain frozen water. If this source of water ice could be confirmed, it would

The Clementine *spacecraft, which orbited the Moon and surveyed its south pole, was part of Phase One of the Moon base project. After orbiting the Moon,* Clementine *continued on to an asteroid, as shown in the upper left of the photograph.*

provide many important uses to a lunar base, including drinking water.

In January 1998, the orbiting *Lunar Prospector* was launched to the Moon to conduct a global mapping survey of the entire lunar surface. Its instruments gave scientists a detailed map of the Moon's mineral composition. The *Lunar Prospector* was also equipped to look for water ice in those permanently shadowed areas near the lunar poles.[4]

In March 1998, the *Lunar Prospector* did indeed find evidence of frozen water on the Moon. It exists in craters at the north and south poles. The ice is mixed with Moon dirt. If the ice can be mined, the water obtained could be used for the Moon base. "This is a significant resource that will enable a modest amount of colonization for centuries," said Dr. William Feldman, an investigator in the study.[5]

The data gathered from robotic probes will influence decisions about where to place the lunar base or bases. It is possible that more than one lunar site will be developed, taking advantage of resources that exist at different locations. Sites will be chosen that will allow the base to become as self-sufficient as possible. For example, a site near one of the lunar poles may be developed so that water ice can be used by the base to provide drinking water and fuel.

Another site might be developed in an area where the minerals in the lunar surface contain higher levels of

In 1998, the Lunar Prospector *found evidence of frozen water at the north and south poles of the Moon.*

oxygen. Processes already exist that can extract the oxygen from the surface. With these processes, the base's occupants could produce considerable amounts of oxygen for the base. The oxygen can be used for breathing. It can also be used as a fuel component for the spacecraft coming and going from the base.

Data from the probes, and the performance of the robotic spacecraft themselves, will also influence the final design of the base's surface systems. These systems include the habitat modules and the base's transportation systems. The data will also affect the design of the spacecraft that will travel back and forth to the Moon.[6]

Eventually the probes will have the entire Moon mapped in great detail. Final choices about the technology that will be the most useful and safe for the base will be made. Toward the end of Phase One, final decisions on lunar sites and Moon base technical systems will be made. The countries involved in the program will agree on their assignments and tasks.

Scientists and engineers will construct new spacecraft and surface modules. Astronauts will begin to train intensely for the new missions scheduled to go to the Moon. The surface modules will be tested on Earth. The new spacecraft will be tested in space to prove their flying and docking abilities.

Phase Two will be ready to begin.

Sometime in the first or second decade of the twenty-first century, the first crew of astronauts will climb aboard their new spacecraft. For the first time in many years, humans will be headed to the Moon.

3

Lunar Outpost

According to NASA's plan, the first surface habitat module will be sent to the Moon without a crew. The habitat will likely be a tube-shaped module. It will be launched from Earth and flown to the Moon by a rocket booster.

Once the habitat module arrives in orbit around the Moon, the rocket booster will separate. The module will be equipped with small rockets and enough fuel to slow its speed and lower itself for a landing. The landing will be managed remotely by controllers on Earth. The controllers will land the habitat module at a site that was determined suitable by earlier probes.

A short time later, the first crew of four astronauts scheduled to land on the Moon will board their spacecraft and prepare for liftoff.[1] This crew of astronauts will be the first to leave Earth orbit since the crews of the Apollo program. Depending on the type of rocket booster used, this first crew's journey to the Moon may take as long as three and a half days or as little as one day.

In the hours before the astronauts enter lunar orbit, the cratered surface of the Moon will begin to fill the view outside their windows. It will no longer appear like a disk to them, as it appears to us on Earth. It will

An automated cargo lander delivers equipment to the surface of the Moon.

become a three-dimensional sphere, with brightly lit fields of craters that fall away into shadows along a curved horizon. It will become for them a desolate but awesome sight to behold.[2]

With controllers monitoring the flight from Earth, the spacecraft will enter orbit around the Moon. Some time later the rocket booster will separate from the lunar landing vehicle, or LLV. Depending on the spacecraft's design, the booster will be either discarded or left in lunar orbit to be reused for the return to Earth.

Following the separation, the LLV will descend toward the lunar surface. Its target will be to land within one kilometer of the surface habitat already in position at the site.[3] If all goes well, the LLV's pilot will use his or her eyes and hands and the spacecraft's onboard flight computer to land the LLV near the habitat. The landing rockets may kick up some lunar dust as the LLV nears the surface. The crew will feel a jolt as the landing legs touch down.

At the moment when that landing is accomplished, astronauts will have returned to the Moon.

There may be a little time for celebrating. However, their landing will be only the first of many steps toward establishing a permanent base on the Moon. The astronauts will be busy right away.

Their landing, and those that will follow, will be scheduled to occur at a time close to what is called lunar dawn.[4] Lunar dawn marks the beginning of a long period

The lunar landing vehicle, carrying a crew of four, descends to the surface of the Moon.

of daylight on the lunar surface. We experience our days and nights on Earth because the Earth rotates, or spins around, on its axis once every twenty-four hours. The Moon has a different rotation period. It takes about a month to rotate once on its axis. This means that astronauts on the Moon will experience a two-week period of day and a two-week period of night.

For this reason it is important that landings to the first lunar outpost be made as near as possible to lunar dawn for that landing site. Landing at this time will give the crews two full weeks of daylight at the beginning of their missions to do explorations and experiments. It is important also because much of their equipment will be powered by solar energy from the sun. Once the sun sets over the lunar horizon, solar energy cannot be generated again for fourteen days.[5]

Shortly after the LLV lands, the crew will help each other get into their space suits and open the airlock to the outside. Each crew member will climb down the ladder to the lunar surface. They will begin unpacking the cargo of supplies and equipment from the lander. Part of the cargo will be a lunar rover, which they will use to carry some of their equipment to the habitat module. Later it will be used to go exploring.[6]

The astronauts will pull out two solar arrays from the LLV to provide power to the lander while its equipment is unpacked. These solar arrays roll out like window shades from either side of the lander. One array will face east and the other west. This will allow the arrays to produce power at all times during the lunar day without having to track the sun as it moves across the lunar sky. With the solar arrays set up in this way, the lander will look as if it is sitting inside a tent.

After unloading the lander, the crew will haul themselves and the equipment to the habitat. Their first tasks will be to open the habitat, climb inside, and begin activating its systems. These systems will be for life support, communications, and living accommodations (such as its bathroom!). Large solar arrays will be rolled out from either side of the habitat to provide power. Once the habitat is inspected and fully powered, the crew will be ready to begin the science and exploration activities they came to do.[7]

During the two-week lunar day, the crew will take turns exploring the area in two-person teams. The first two astronauts will slide into their space suits and venture outside through the habitat's airlock. Both will carry equipment they will need for their work. The work

The lunar habitat will contain life support and communications systems.

done outside the habitat in space suits is called extravehicular activity, or EVA.

The partners will load their equipment aboard the lunar rover. They will place themselves and their bulky life packs on the seats and drive out over the dusty landscape of the Moon. They will perform a wide range of tasks: They will make geologic studies of the lunar surface, bringing samples back to the habitat for analysis. Experimental equipment will be set up to make studies of particles reaching the lunar surface from the sun. Astronomical telescopes will be assembled and anchored in place. The astronauts will also test new equipment that scientists hope can be used in mining the lunar surface for essential elements such as oxygen and silicon.[8]

When astronauts return to the Moon to begin construction of the Moon base, they will, of course, be wearing space suits during their work. The rest of the time they will live inside pressurized lab or habitat modules. The modules are needed because the Moon has no atmosphere.

Without an atmosphere, the Moon is exposed directly to the sun's cosmic rays. Some of these rays may be harmful. Astronauts should be protected from them while in their space suits or habitat modules. Scientists will still be looking for any harmful effects caused by living on the Moon for extended periods, such as overexposure to these cosmic rays.

Lunar rovers allow astronauts to travel far from their habitat module.

Astronauts on these first missions to the lunar outpost will conduct about thirty EVAs. Because of the effort involved in getting in and out of the space suits and the habitat airlock, each EVA will last about ten hours.[9] Ten hours is a long time to be in a space suit. For that reason, the suits will be equipped with a system that will allow the astronauts to go to the bathroom while in the suit. The helmet's interior will include a small hose that the astronaut can use like a straw to drink liquid from a pouch in the suit. A special fruit-and-nut bar can also be placed in a pouch inside the helmet so that the astronaut can turn and eat it if he or she becomes hungry.[10]

While one pair of astronauts is outside on the lunar surface, the other pair will remain in the habitat to conduct additional experiments and tests. Both pairs of astronauts will be trained in conducting EVAs. The following day, the second pair will go outside, and the first pair will remain in the habitat to conduct experiments. After an astronaut pair conducts a full EVA, it will not conduct another EVA for two days. This schedule will prevent the astronauts from becoming too tired.

The lunar rover, or Moon buggy, will be powered by an electrical battery, charged from a solar array. While astronauts drive the rover, the battery will provide its power. Whenever it is parked, the rover's solar array will generate energy and feed it to the battery until it is fully charged again.

The rover will also be equipped with video cameras and a remote driving system. By using this system, crew members will be able to operate the rover from inside the habitat if necessary. This will allow astronauts to send samples from the work site to the habitat while the two EVA astronauts continue working at the site. The rover could also be sent to pick up an astronaut whose explorations have carried him or her a long walking distance from the base.[11]

After six days of their busy work routine, the weary

crew members will earn a day off. This will be a time for them to rest and relax in the habitat, talk to their families by satellite, or simply look out the habitat's windows at the desolate lunar landscape.

After the astronauts have taken turns inside and outside the habitat for two weeks, the sun will set over the lunar horizon. Lunar night will begin.

During a stay at the Moon base, astronauts will have some free time to enjoy the landscape of the Moon.

For the next fourteen days, light conditions at the lunar outpost will be very dim. The small amount of available light will actually be sunlight reflected from Earth. During this period, EVAs will be limited to times when the lighting conditions are sufficient. These times will most likely be in the days immediately after lunar sunset and immediately before lunar dawn. Most of the crew's activities during lunar night will be focused on laboratory work within the surface habitat.

Lunar night will also be an ideal time for the astronauts to conduct astronomical observations with optical and radio telescopes. During lunar night there is no direct sunlight to obscure observations of the stars and planets. There is also no interference from human-generated radio waves on Earth. For these astronomical observations, the far side of the Moon is the place to be.

After fourteen days of darkness, the sun will rise once more above the lunar horizon. The astronauts will return to their intense work schedule of EVAs and laboratory work within the habitat. It will have become a busy but normal routine for the four crew members. Their valuable experience with the systems and equipment of the lunar outpost will benefit the crews that follow them. The knowledge and information they will have gained about the site may already have influenced decisions about what equipment will be sent next to expand the outpost.[12]

The crew will spend about forty days at the lunar

outpost before they begin packing for the return to Earth. The systems of the surface habitat will have to be carefully shut down and prepared for the arrival of the next lunar crew.[13] The astronauts will make several trips between the habitat and the lander with the rover, carrying loads of sediment and rock samples, specimens, and results from experiments.

The solar arrays will then be rolled up and the cargo bays will be locked. When everything is packed and secured, the crew will climb aboard the lander and seal the hatch behind them. The systems of the LLV will be

In preparation for their return to Earth, the Moon base crew refuels the propellant tanks of the spacecraft.

brought up to full power in preparation for liftoff from the Moon.

The crew will strap themselves in and listen as the countdown progresses.

"T minus one minute until lunar liftoff," Mission Control will say.

"All LLV ascent systems are go," the crew commander will reply. The crew will then wait through the final seconds.

"Ten, nine, eight, seven, six, five, four, three, two, one . . . Liftoff!"

The crew will suddenly feel themselves pressed back in their seats as the LLV launches them up and away from the lunar surface. The LLV will carry them into lunar orbit where they will likely dock with a return-flight rocket booster. When the booster fires to break them out of lunar orbit, the first crew to visit the lunar outpost will be on their way home.

The Moon will not be left alone for long. Only a few weeks or months will pass before the next crew of astronauts will arrive at the lunar outpost. More experiments will be conducted and more explorations will be made. New equipment will be delivered to the site, and new technologies will be tested to see how they might function in the operation of a larger Moon base.

The outpost will progress gradually but steadily from this point. Eventually a second habitat module may be landed near the site, and possibly a third. In time, the flights to the outpost will become so regular and frequent that the outpost will be occupied by at least one crew at all times. Later, equipment for mining and processing the lunar sediment for needed resources will be landed near the outpost.

The transition from Phase Two to Phase Three will begin. Over a period of years, the site will gradually become something more than a lunar outpost. It will become a busy, working Moon base.

4

A Busy Base

The Moon, which seems to be a quiet ball in the night sky, may be a very busy place by the year 2030. By that time people, supplies, and equipment should be traveling to and from the Moon on a regular schedule. The reason for this traffic around the Moon is that a lot may be happening at the Moon base.

At this stage, the Moon base will no longer be a simple scientific outpost. It will be a center of lunar science and industry. Sometime between 2020 and 2030, the Moon base will likely consist of a number of different work and habitat modules.[1]

Engineering tests and scientific experiments will still be conducted at and around the site. But by this time they will no longer be the sole purpose and function of the Moon base.

A facility will have been constructed near the base to process lunar sediment from the surrounding craters and hills.[2] The process will release oxygen molecules trapped in the sediment. The oxygen will then be collected and stored in large tanks.

The oxygen will be used for two major purposes. The most immediate purpose will be to provide quantities of breathing oxygen for the occupants of the Moon base. The second major use will be as a component of the propellant needed for spacecraft arriving at and

The Moon base may someday be a busy community with different work and habitat modules.

departing from the base.[3] With this oxygen available, oxygen supplies will not need to be carried to the Moon from Earth.

Lunar-derived oxygen is called LUNOX. Production of LUNOX will greatly decrease the cost of traveling to the Moon. The most expensive part of space travel is the cost of launching heavy objects out of Earth's gravity. Carrying that oxygen to the Moon would mean that even *more fuel* would be needed to launch the added weight of extra oxygen out of Earth's gravity.

With the LUNOX facility working at the Moon base, spacecraft will be resupplied with LUNOX once they reach the Moon. Lifting the LUNOX off the Moon will be done at only a fraction of the cost of launching and carrying the liquid oxygen out of Earth's much stronger gravity. It will be a cost savings of hundreds of millions of dollars.[4]

The LUNOX facility will not be the only industry at the Moon base. After oxygen, the most abundant element in the Moon's crust is silicon, and this element has a very important use in space. Silicon is the element used to create solar power arrays. Solar power arrays, which use energy from the sun to produce electrical power, will be used for power at the Moon base, on the International Space Station, and on nearly every spacecraft flying between Earth and the Moon and outward to other planets.[5]

Oxygen will be mined on the Moon at the LUNOX production facility.

The Moon base will be able to process the Moon's silicon into small solar panels. Another facility, or factory, will assemble the panels into different solar array designs. These will be used for power generation at other lunar outposts or on spacecraft awaiting delivery of the arrays from the base. It is possible that the solar arrays will have tiny labels that read MADE AT MOON BASE or PRODUCT OF THE MOON.

By about 2030, the Moon base will be a busy place of industry whose resources will enable humans to reach out to other planets in the solar system. It will prepare the way for humans to go to Mars and will transform space travel into a successful and far less costly system than it is today.

The Moon base will teach us what it is like to work on the surface of an alien world day after day. Some people on the Moon base will need to get into space suits almost every day to go out onto the surface and operate equipment. They may work at a LUNOX facility or at a silicon-processing plant. Others who go out onto the surface regularly may be technicians who repair equipment that has failed.

These jobs will no doubt be difficult at times. The workers will probably spend long hours inside the suits. It may be frustrating at times for them to use their tools or get into cramped places because of the bulkiness of the suits. But the suits will be bulky for good reasons.

The complex fabric of the space suit must be strong enough not to tear if it gets caught on a piece of equipment. It must also protect the workers on the surface of the Moon from micrometeorites.[6] Micrometeorites are any small space debris, including dust particles, that can hit the Moon at very high speeds. These collisions happen because the Moon lacks any atmosphere to burn up these particles before they hit the surface—or anyone who happens to be walking on it.

Other parts of their job will be amusing. With only one sixth the gravity of Earth, the Moon will be a fun place. In fact, whenever the colonists walk from one area to another, they will not exactly *walk*. They will more likely move in a motion something like a skip or a hop. With little effort, they will be able to bounce gently along the Moon's dusty surface as if they were on a trampoline.[7]

One astronaut might say to another, "Let's hop over to the LUNOX facility." And that is exactly what they would do—hop all the way to the LUNOX facility.

Others on the Moon base will spend most or all of

Solar panels, like the blue one seen here on the Sojourner *rover, may someday be manufactured on the Moon.*

their time inside the pressurized modules. Some of them will work in laboratories; others will monitor the various mining operations for LUNOX and silicon. Spacecraft will come and go regularly at the Moon base. Eventually a number of people and loads of equipment will arrive for various training operations for a future trip to Mars. The Moon base will provide the ideal training ground for the people expected to make that journey.

Spacecraft designed to land on Mars can be tested on the Moon's surface. The astronauts expected to go to Mars can rehearse their operations and practice using their equipment. They can even spend weeks or months at a time at a simulated Mars site on the Moon.[8]

The Moon base will be a busy place. But no matter what the people at the base will be doing, it is possible that many of them will stop to watch something that happens on the Moon only once every twenty-eight days: the rising of Earth above the lunar horizon.

Seeing the blue Earth rising at one end of the lunar sky and the bright stars shining at the other might make a vivid picture in the minds of those astronauts who are training on the Moon for a trip to Mars. To look from one end of that sky to the other might be an inspiring reminder to them of where they came from and where they are going.

The Moon base will be a necessary step in learning how to live away from Earth. Like Neil Armstrong's first steps upon the lunar surface, a base on the Moon is a small step that will lead to giant leaps.

No matter how long it takes, those giant leaps will happen. Someday, farther into the future, the work and exploration accomplished on the Moon base may lead to human beings living on other worlds.

The rising Earth, seen from the Moon, will be an inspiring sight to astronauts working at the Moon base.

CHAPTER NOTES

Chapter 1. One Small Step

1. *Apollo 11, Technical Air-to-Ground Voice Transcription*, Manned Spacecraft Center, Houston, Texas, July 1969. All in-flight communications in this chapter come from this source.

2. Ibid.

3. Wendell W. Mendell, "Lunar Base as a Precursor to Mars Exploration and Settlement," Solar System Exploration Division, NASA, Johnson Space Center, Houston, 1990, p. 1.

Chapter 2. The Next Step

1. Bret G. Drake, "Alternative Lunar Mission Strategies," NASA, Johnson Space Center, Houston, 1991, p. 2.

2. Inge Sellevag, "Back to the Moon the European Way," *Ad Astra*, May/June 1995, p. 35.

3. Drake, p. 2.

4. Pat Dasch, "Lunar Prospector: NASA's First Lunar Mission in 23 Years," *Ad Astra*, May/June 1995, pp. 31–32.

5. Warren E. Leary, "Spacecraft Sees Signs of Large Amounts of Water Ice in Scattered Craters on the Moon," *The New York Times*, March 6, 1998, p. A16.

6. Drake, p. 2.

Chapter 3. Lunar Outpost

1. P. D. Campbell, "First Lunar Outpost Surface Habitation Phase Crew Time Analysis," prepared by Lockheed Engineering and Science Company for Flight Crew Support Division, NASA, Johnson Space Center, Houston, December 1992, p. 2.

2. *Apollo 11, Technical Air-to-Ground Voice Transcription,* Manned Spacecraft Center, Houston, Texas, July 1969.

3. Campbell, p. 5.

4. Ibid., pp. 3–4.

5. "Photovoltaic Power for the Moon," *Canadian Space Gazette*, 1996, pp. 1–2.

6. Campbell, p. 4.

7. "Photovoltaic Power," pp. 1–2.

8. Bret G. Drake, "Alternative Lunar Mission Strategies," NASA, Johnson Space Center, Houston, 1992, p. 2.

9. Campbell, pp. 4–5.

10. PBS Video, *Astronauts*, Carlton U.K. Productions, 1996.

11. "Photovoltaic Power," pp. 1–2.

12. Drake, p. 2.

13. Campbell, p. 4.

Chapter 4. A Busy Base

1. Stanley K. Borowski, "24-Hour Commuter Flights to the Moon Using Nuclear Rockets with LUNOX Afterburners," *Ad Astra*, July/August 1997, pp. 24–25, 28.

2. Carl Allen, "Prospecting for Oxygen on the Moon," *Ad Astra*, November/December 1996, pp. 34–35.

3. Ibid., p. 34.

4. Ibid.

5. "Photovoltaic Power for the Moon," *Canadian Space Gazette*, 1996, pp. 1–2.

6. Frank Kuznik, "Spacesuit Saga: A Story in Many Parts," *Air & Space*, September 1997, pp. 42–43.

7. John and Nancy Dewaard, *NASA: America's Voyage to the Stars* (Greenwich, Conn.: Brompton Books Corp., 1990), p. 66.

8. Wendell W. Mendell, "Lunar Base as a Precursor to Mars Exploration and Settlement," Solar System Exploration Division, NASA, Johnson Space Center, Houston, 1990, pp. 7–10.

GLOSSARY

capcom (capsule communicator)—Person at Mission Control who is primarily responsible for communicating with astronauts during a mission.

docking—Joining two spacecraft in space.

extravehicular activity (EVA)—Any human space activity that takes place outside the crew compartments of spacecraft or surface habitats.

lunar dawn—The period when the sun begins to rise over the lunar horizon, beginning the two-week period of lunar day.

lunar day—The two-week period of daylight on the Moon. Daylight lasts this long because the Moon takes nearly a month to rotate once on its axis.

lunar landing vehicle (LLV)—Spacecraft, either controlled remotely or piloted, that are designed to land on the lunar surface.

lunar night—The two-week period of darkness on the Moon. Night lasts this long because the Moon takes nearly a month to rotate once on its axis.

LUNOX (lunar oxygen)—Oxygen derived from the Moon by chemically processing oxygen-rich lunar sediment.

Mir—The world's first permanent space station launched by Russia in February 1986. Through the late 1990s, cosmonauts and astronauts continued to man this experimental laboratory in space.

NASA—The National Aeronautics and Space Administration. This is the United States government agency that administers the country's space program.

rover—A small vehicle that can be driven by astronauts or controlled remotely by radio.

Salyut—The first manned space station. Russian cosmonauts boarded the station in June 1971 for the 23-day mission. The mission ended tragically when all three men died on the return flight when their spacecraft cabin lost all air pressure.

solar array—Silicon panels that use light from the sun to generate electrical power.

space station—A permanent laboratory orbiting Earth that allows astronauts to conduct long-term research.

Skylab—The first United States space station, launched in May 1973. It fell out of orbit in July 1979 and fell apart. Parts landed in Australia and in the Indian Ocean.

FURTHER READING

Books

Baker, David. *Living on the Moon.* New York: Rourke Book Company, Inc., 1989.

Burgess, Eric. *Outpost on Apollo's Moon.* New York: Columbia University Press, 1993.

Cole, Michael D. *Apollo 13: Space Emergency.* Springfield, N.J.: Enslow Publishers, Inc., 1995.

Dudley, Susan. *Countdown to the Moon.* New York: Crestwood House, 1992.

Taylor, Glenn. *A Kid's Guide to Living on the Moon.* New York: Capstone Press, 1991.

Internet Addresses

JSC/NASA. "Lunar Documents." *Planetary Missions and Materials Exploration Server.* May 30, 1996. <http://www-sn.jsc.gov/explore/Data/Lib/DOCS/EIC040.HTML> (July 16, 1998).

Prado, Mark. "Lunar Material Utilization." *PERMANENT.* March 1, 1998. <http://www.permanent.com/1_index.htm> (July 16, 1998).

The University of Arizona/NASA. "Research and Development Projects." *Space Engineering Center for Utilization of Local Planetary Resources.* n.d. <http://scorpio.aml.arizona.edu/projects.html> (July 16, 1998).

INDEX